Ford Goes Looking for the Sun

by Joseph Settecasi

illustrated by Lisa Bohart

I0034319

Copyright 2017 © Joseph Settecasi
All rights reserved.

No part of this publication may be reproduced,
stored in a retrieval system, transmitted in any form
or by any means, electronic, mechanical, photocopying,
recording, or otherwise, without prior written
permission of the publisher and author/illustrator.

Printed in the U.S.A. by Ingram, Nashville, TN

Ford Goes Looking for the Sun
Authored by Joseph Settecasi
Edited by Jane Brandi Johnson
Graphic design and illustrations by Lisa Bohart
Pre-press production by Sven M. Dolling

ISBN-13: 978-0-692-96833-8
LCCN: 2017916070

Printed in October 2017

This book is dedicated to my grandson Ford, whose
sense of life and wonder is the inspiration for this story,
and to all the teachers and parents who explain the
wonders of our world to their children each day.

Hidden Words and Symbols

Within each two-page spread is a 'hidden word or symbol' for children to find.

These words and symbols are science concepts which relate directly to the story.
More importantly, they provide scientific information that can be further studied and
explored with a parent or teacher.

A key to the hidden words and symbols can be found at the back of the book.

Ford Goes Looking for the Sun is about us
and how we fit into the world around us.

Joseph Settecasi

Everyone knew it. Everyone knew that he was the most curious boy they had ever met. He wanted to know **EVERYTHING!**

So Ford, the red-headed boy with the fire and passion of the Sun, begins a journey to find the Sun.

His mission is to ask the Sun a question.

Not far from home, Ford arrives at a big open field, looks around, and sees something big and blue.

"Hello Sky," says Ford.

"Do you know me?" asks the Sky.

"Yes," says Ford. "You are made of the air I breathe every day."

"Sky, have you seen the Sun?"

"Not today," says the Sky, "but the Sun is my friend because its light makes me bright and blue."

"Nice seeing you, Sky," says Ford.
"I'll keep looking."

Ford keeps walking and comes to a gurgling, flowing creek.

"Hello Water," says Ford.

"Do you know me?" asks the Water.

"Yes," says Ford. "I drink you every day."

"Water, have you seen the Sun?"

H₂O

"Not today," says the Water, "but the Sun is my friend because its heat turns me into clouds and rain."

"Glad to see you, Water," says Ford. "I'll keep looking."

Close by, Ford sees something tall and branching, with brown bark and big green leaves.

"Hello, Mr. Tree," says Ford.

"Do you know me?" asks the Tree.

"Yes, you are a mighty Tree," says Ford. "Mr. Tree, have you seen the Sun?"

"Not today," says the Tree, "but I can tell you that the Sun is my friend. The Sun shines on me so I can grow to make the fruit and nuts you eat every day."

"Great to talk to you, Mr. Tree," says Ford. "I'll keep looking."

Ford continues on and finds a big pond with something swimming in it.

"Hello Fish," says Ford.

"Do you know me?" asks the Fish.

"Yes, you are a Perch," says Ford. "This pond is your home. Fish, have you seen the Sun?"

"No, not today," says the Fish, "but the Sun makes plants grow so that I have something to eat. The Sun is my good friend."

"Great to talk to you, Fish," says Ford. "I'll keep looking."

On a rock at the edge of the pond, Ford sees something leaping and hopping. "Hello Frog," says Ford.

"Do you know me?" asks the Frog.

"Well, yes. You are a freshwater Frog," says Ford. "You live in this pond with the Fish. Frog, have you seen the Sun today?"

"No, not today," says the Frog, "but the Sun is our good friend. The Sun's heat warms us so we can swim and play."

"Have fun, Frog," says Ford. "I'll keep looking."

MAMMAL

Ford keeps walking and soon meets a big, brown, furry creature eating an apple from a tree.

"Hi there, Bear!" says Ford.

"Do you know me?" asks the Bear.

"Yes, you are a Brown Bear," says Ford. "You live in the woods with all the other animals. Bear, can you help me find something?" Ford asks.

"Yes!" says the Bear. "What are you looking for?"

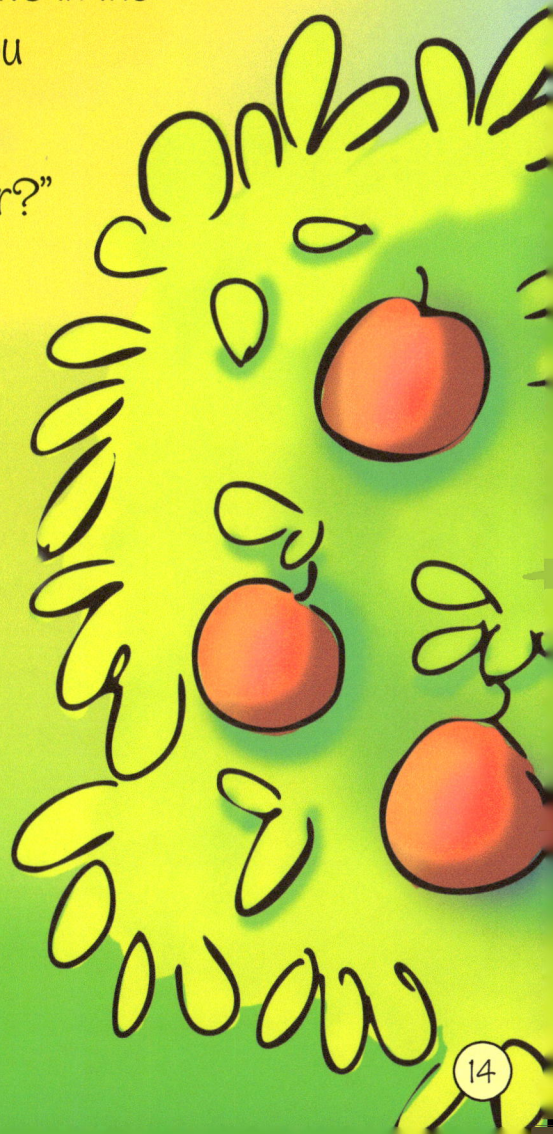

"I've been looking for the Sun for so long," says Ford.

"Bear, do you know where I can find the Sun?"

"Sure," says the Bear. "The Sun just came out from behind the clouds."

"There's the Sun!"

STAR

Ford looks up and sees a BRIGHT-YELLOW BALL
in the Sky. The Sun is powerful and too bright
to look at, so Ford turns his eyes away and says,

"Hello Sun!"

"Do you know me?" asks the Sun.

"Yes, *everybody* knows you!" says Ford with
a smile. "I've been looking for you," says Ford.
"May I ask you a question?"

"SURE!" says the Sun.

RAYS

20

"What are you?" Ford asks.

"I AM ENERGY.
I SHINE OUT LIGHT
AND HEAT, SO THAT ALL
CREATURES ON EARTH CAN LIVE."

Ford thanks the Sun. **"You're great!"** he says.

On his way back home, Ford thinks about all the wonderful things he met on his journey to find the Sun.

Back at home, his Mom asks, "Ford, did you find the Sun?"

"Yes," says Ford, "and the Sky, the Water, the Tree, the Fish, the Frog, and the Bear."

But Ford's Mom knows him well.
The Sun had answered his question,
but she still sees that curious look on his face!

"What is it, Ford?" she asks.

"I have one more question," he says.

"Mom, now I know the Sun, the Sky, the Water, the Tree, the Fish, the Frog, and the Bear.

But – *What am I?*"

"Ford, there is something wonderful inside all things on Earth. It is inside you, inside me, and inside every living creature. All things on Earth are connected because we are all made of the same stuff.

And... what are *you*, you ask?"

"YOU ARE EVERYTHING IN THE WORLD, MY SON.
EVERYTHING IN THE WORLD!"

Key to Hidden Words and Symbols

"The Sun shines not on us but in us."

John Muir

www.ingramcontent.com/pod-product-compliance
Lightning Source LLC
Chambersburg PA
CBHW041724210326
41598CB00007B/767